A brief history of the world – The great deception

From alpha to omega

Eternity begins when the duality of light and darkness returns to its original source

Dr. Ali Ansarifar

Kingdom Publishers

Copyright© Dr Ali Ansarifar 2024

All rights reserved. No part of this book may be reproduced in any form by photocopying or any electronic or mechanical means, including information storage or retrieval systems, without permission in writing from both the copyright owner and the publisher of the book. The right of Dr Ali Ansarifar to be identified as the author of this work has been asserted by him in accordance with the Copyright, Designs and Patents Act 1988 and any subsequent amendments thereto.

A catalogue record for this book is available from the British Library.

All scripture quotations have been taken from the Interlinear Bible

ISBN: 978-1-916801-01-1

1st Edition 2024 by Kingdom Publishers, London, UK.

You can purchase copies of this book from any leading bookstore or email **contact@kingdompublishers.co.uk**

Disclaimer

It has never been the intention of the author of this book to infringe on the sensitivity, personal faith, or beliefs of his readers. The material and information presented in this book are never meant to cause offence to any individual person or group of people of any race, nationality, religion, or background. This book is written for educational purposes and academic interest only.

Dedication

I dedicate this book to my father and all those who supported me during my Christian journey.

Foreword to Ali Ansarifar's book series

If you were asked to write a history of our universe and humankind's place in that story, where would you begin? An astrophysicist, I imagine would start with what science tells us. Dr. David Wilkinson, a respected cosmologist, also a theologian, writes that current scientific theories describe the universe back to a time some 15 billion years ago when it was a tiny, tiny fraction of a second old.

However, this isn't the starting point for Dr. Ali Ansarifar. In fact, he doesn't address the cosmic issue, our planet, and the massive diversity of its life forms until further on in the series, when he gives a concise account of where modern astrophysics and evolutionary science stand on the issue.

Ali Ansarifar's primary purpose in this eight-book series is an exploration of what theologians know as Salvation History. In other words, explaining the story of our world and its human race in terms of meaning and purpose, theologically speaking, teleology history, whether ours or the world's could be seen as a mere list (a very long one indeed) of events, facts, and figures, the rise and fall of nation states and empires, wars, and peace; the advance of science; but without seeking some purpose to it all would be an extremely tedious and boring way of telling a story.

Dr. Ansarifar's concern in this series is "What does it all mean?" How can that ancient duality, the 'light and darkness, good and evil paradox, be resolved. At times, he deliberately stands outside of Judeo-Christian belief. In other words, to a non-believer, billions of years ago, there was an inexplicable flash of light that heralded not just the beginning of creation as we understand it, but also that dualism, light and darkness. In 1916, an agitated Henry Ford, the industrialist, proclaimed, "History is more or less bunk," he was later to modify this by explaining how to him, conventional historical accounts seem to be full of military heroes and politicians. In truth, our human story on this planet is multifaceted. To understand what it's all about, one must view it from many angles, something akin to those online 'virtual tours' of historic buildings, even the hotel you wish to book for a holiday! Dr. Ansarifar has approached the task of trying to make sense of our human story, "the good, the bad, and the ugly" by including a brief excursion into the world faiths, including his own cultural background, because, like it or not, the tide of secularism hasn't been able to wash away a deep-rooted feeling. Most people agree that our existence has little meaning and purpose without that 'something' or 'someone' beyond what science can tell us.

Most of Ali Ansarifar's time and writing are concerned, however, with our Judeo-Christian heritage, viewed historically and with a look ahead as to how it all might turn out. Of course, the vast field of 'End Times' speculation in Judeo-Christian writing is centuries old, Jewish Apocalyptic, and Christian attempts at understanding its significance for our present times are as varied as British weather. However, whoever delves into what the future may hold should bear in mind that Apocalyptic writers were primarily concerned with their own near future and not events up to two millennia ahead. So, for us, all such speculation must come with a health warning. Dr. Ansarifar offers several scenarios, leaving readers to evaluate and make up their own minds, for instance, whether Jewish Zionists would ever getting a Third Temple erected on Temple Mount seems highly unlikely in the foreseeable future, given the volatility of Middle-East politics and what would be massive international opposition. However, Dr. Ansarifar invites us all to at least consider possibilities in the near and distant future and, laying aside the many interpretations of John's Apocalypse, that book's message can be summed up succinctly in the words of the 14th Christian mystic, the first known woman mystic in fact, Julian of Norwich said, "I shall make all things well, and you shall see for yourself, all manner of things shall be well." God, as revealed to us through His Son, Jesus Christ, shall triumph. His light was never and shall never be extinguished.

Rev Martin Mitchell

Baptist/Methodist Minister, former Lecturer in World Faiths, and Presenter at Premier Radio.

Scraptoft Teacher Training College, Leicester, United Kingdom.

From alpha to omega

At the beginning, there was darkness, and then light came into existence of its own accord. This was the beginning of duality. Duality persisted in the light in the form of deficiency, which it inherited from darkness. To redeem itself, light (God) created humankind in his image and likeness. Humanity was marred by idolatry, moral decay, and corruption but strived to redeem itself when faith in the inner heart of man led him to long for an invisible God. This was the Abrahamic faith. The Abrahamic faith struggled with immorality and could not free itself from the duality of light and darkness. So, Jesus was conceived with the Abrahamic faith, and God's moral law was written on his heart to end duality. Jesus is the perfect oneness through whom both God and humanity will be redeemed from duality and freed from its destructive power. But for now, we are captives to this merciless duality.

The history of the world hinges around two main periods: the rise of empires and, more recently, the Christian era. Empires have risen and collapsed cyclically, leaving behind devastation, ruins, and shattered dreams for millions of their subjects. The destiny of humanity has been intermingled with the rise and fall of empires. God Christianised the pagan Roman empire before it collapsed to create the Papacy and bring the reign of the empires to an end. The Papacy faced a major challenge to its legitimacy when the Protestant Reformation identified its popes as the line of antichrist. The Counter-Reformation devised a method to remove this challenge to the Catholic Church through deception, and the rebuilding of the Third Temple in Jerusalem is at the centre of this plan. But the Jewish return to the holy land and the rebuilding of the Third Temple in Jerusalem are not included in the Bible or in the end-time prophecies of Jesus Christ. The main figure is the man of sin, or antichrist, who will sit in the Temple and claim to be God. Jesus said, "See that not any leads you astray." For many will come in My name, saying, "I am the Christ." (Matt 24:45). The mysteries of God and the creation of humanity are revealed on the cross of crucifixion. Humankind's fate is tied to the cross until the end of time.

The gospels according to Philip and Thomas are not included in the Bible. They are referred to as apocrypha. They contain wisdom and sayings that are truly unique. Information in these gospels will aid in understanding Jesus' profound and sacred teachings to his disciples. We may also benefit from this treasure trove of wisdom.

This is a story like no other.

Contents

Chapter 1
When it all began – Light and darkness coexisting ... 26

Chapter 2
An empire ending all empires – End of history ... 45

Chapter 3
The great deception – Israel in the Holy Land ... 54

Chapter 4
Summary and conclusions ... 65

Chapter 5
Commentary on the Apocrypha of Philip and Thomas ... 85

The three pillars of our existence

The scheme of Providence ... 93

References ... 95

Afterword ... 100

About the author ... 101

About the book ... 102

Final notes by the author ... 103

Books by the author ... 104

A brief history of the world

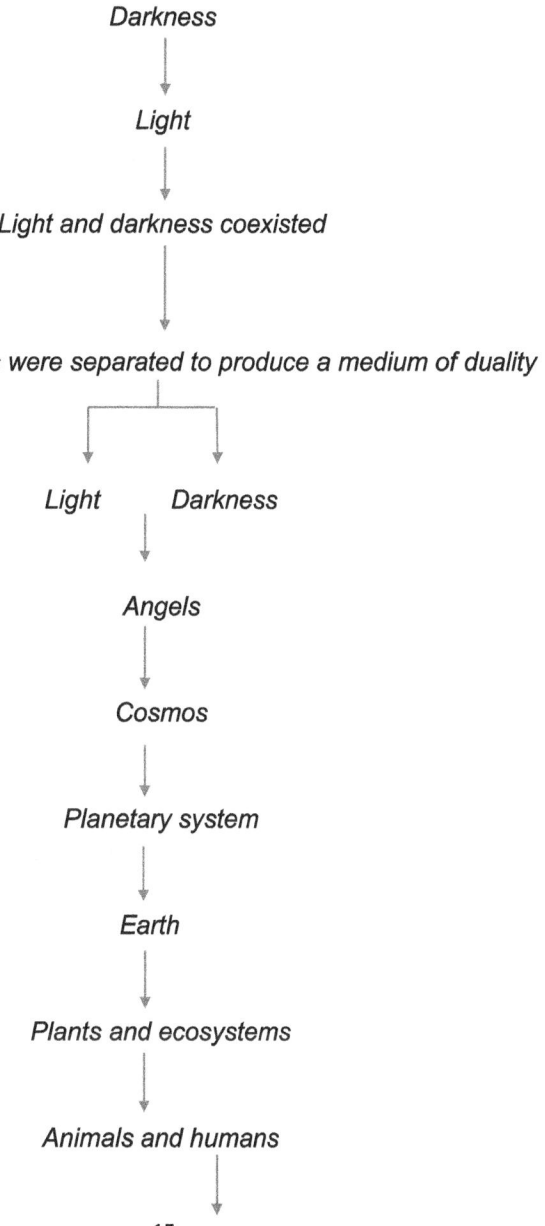

Primitive
↓
Anthropological
↓
Religious
↓
Spiritual
↓
Abrahamic faith
↓
Dissolution of duality in Jesus Christ
↓
Christianisation of the pagan Roman Empire
↓
Western European Enlightenment
↓
Judaic world order and the rise of the man of sin
↓
The return of Christ and the destruction of the man of sin
↓
The dissolution of duality in its entirety
↓
God's kingdom and the kingdom of darkness
↓
Eternity begins

The Disciples said to Jesus, "Tell us how our end will come to pass." Jesus said, "Then have you laid bare the beginning, so that you are seeking the end? For the end will be where the beginning is. Blessed is the person who stands at rest in the beginning. And that person will be acquainted with the end and will not taste death."

[1] (Thomas 18)

So, let us visit the beginning, where it all happened, to understand the end.

God said, "Let light be" — and there was light. And God saw the light, that it was good, and God separated between the light and the darkness. (Genesis 1: 3-4)

Duality arose when light emerged of its own accord from darkness. And from the duality came the world's struggle. It is only in the person of Jesus Christ that duality is nullified, and perfect oneness is achieved. Happy are those who make the journey to Christ and leave behind the destruction and mayhem of this world to find rest for their souls.

There is peace and fulfilment in the oneness of Jesus Christ. And there is confusion and damnation in the duality of things. God's mercy reaches those who choose to leave the duality of the world behind through repentance and seek salvation in Jesus Christ. It is the longing of the inner heart for the divine that frees a mortal from the clutches of duality, and when the inner heart is coupled with God's moral law, perfect unity is achieved, like in Jesus Christ.

To deceive and enslave humans, the forces of evil and their principalities wrapped the biblical truth in lies, pretends, claims, and counter-claims and presented it to the world as the truth. It is in Jesus Christ that this deception is unmasked, the truth is revealed, and the righteous man is freed from its clutches.

The truth is revealed to the meek and righteous without theatrics. It is at the heart of the coming deception to hypnotise the senses and seduce humans into believing lies and acting ungodly. What a terrible fate awaits those who will be deceived!

Light has constructive qualities such as creativity, decency, compassion, forgiveness, healing, wisdom, and a good life. There is destructiveness, hatred, lies, confusion, dishonesty, ignorance, and death in the darkness. It is puzzling why so many humans choose darkness over light. Then again, light knows his children, and darkness knows hers.

The faithful's inner hearts yearn for God, but they struggle with the keeping of the moral law and commit sins, alienating them from God. God removed alienation by writing His moral law on Jesus' heart, who had faith like Abraham, and produced a perfect being who is freed from the clutches of destructive duality for all time. Now all the faithful can be made perfect like Jesus Christ and inherit eternal life.

Why have the gentile and Jewish worlds merged since the Babylonian invasion of the Holy Land?

Imagine a flooded plain where a stream appears. The water pours into the stream and flows to an end, where it will be disposed of. Now imagine a world full of sin and corruption, which will manifest itself as the man of sin before the second coming of Christ. What should be the stream into which the dark forces of evil must flow to their end?

The stream is running in front of our eyes towards its dreadful end, but we do not comprehend it!

This is where the greatest deception in human history is taking shape.

Chapter 1

When it all began – Light and darkness coexisting

Light came into existence of its own accord from darkness. Light and darkness co-existed, and this was the beginning of duality. Duality persisted in the light (God) in the form of deficiency, and God decided to remedy the duality. To understand and remove the deficiency, God created humanity in His image and likeness. However, creation was corrupted by the deficiency in God at every stage, causing angelic rebellion, cosmic chaos, planetary destruction, the extinction of life, human rebellion, and sins. Humanity's deficiency strives to be remedied through faith in the inner heart of man and in the invisible God. This was the beginning of redemption. The faith of the inner heart was coupled with God's moral law in the person of Jesus Christ to bring about perfect unity, which ends the deficiency in both man and God. The duality will be dispersed to its original source after humanity is redeemed in Jesus Christ, and God will be freed from deficiency to live as a perfect being in the redeemed humanity and kingdom of pure light.

Creation of angels – God separated the light from the darkness. And God called the light "Day." And he called the darkness "Night." (Genesis 1-5) Light and darkness were one before God separated them. But how can light and darkness be one? Darkness existed, and light, of its own accord, came into existence from the darkness and stood at rest. [1] Light (God) inherited duality from its source, darkness, resulting in a deficiency in Him. For example, God unleashes his wrath on unbelievers (John 3:36), kills (1Samuel 2:6), commands others to kill (Deut 20:16-17), curses (Malachi 2:2), and wages war (Jeremiah 51:20-23). God also blesses (Deut 21:1-5), shows compassion and forgives sins (Hebrew 8:12), heals the broken-hearted and binds up sorrows (Psalm 147:3), grieves and shows remorse (Genesis 6:5-6), grants mercy to sinners (Zechariah 7:8-10), and is gracious beyond measure (Ezra 9:8). Clearly, the deficiency resulting from duality is affecting God's conduct adversely. God wanted to remove the duality in him and gain perfect oneness with no deficiency, so he created humankind in his own image and likeness (Genesis 1:27). However, before God made the heavens, the earth, and humanity, He created angels. When the angels were created from light, Lucifer, the son of the morning, desired to ascend into heaven, exalt his throne above God's stars, and become like God. He assembled some angels of heaven and fought for supremacy. The war in heaven began, and Lucifer, known subsequently as Satan, and his angels were defeated and cast down to the earth in dishonour. Another group of angels, using the image of God as a template, made a man in their likeness and called him Adam (man). The Watchers, the Lord's angels, were tasked with teaching the children of men how to be just and upright on earth. But the Watchers lusted after the daughters of men, killed men, took their daughters for themselves, and had children with them. Their children were giants in the earth and performed every abhorrent deed, blasphemous utterance, and sexual immorality. [2] The angelic world became arrogant, disobedient, rebellious, and warlike. God is the source of all that is creative and constructive, but all his creative work has been spoiled by deficiency at every stage. After the angels were created, God made the cosmos.

Creation of the cosmos – The cosmos was created by a violent event in the past referred to as the "Big Bang." Some 12 to 20 billion years ago. After millions of years, the universe was stuffed with large, irregular aggregations of gas or stars held loosely together by the force of gravity. The modern galaxies were emerged from these cosmic chrysalises, called protogalaxies. There are an estimated 200 billion galaxies in the observable universe, with each galaxy averaging 100 million stars. [3] This suggested that the universe had a beginning and was expanding. All the galaxies were rushing away from each other, increasing the space between themselves and their neighbours. At the beginning, all observed matter was compressed into a dense plasma whose temperature should have been 10 billion degrees on the Kelvin scale. As the universe cooled, matter was formed in the fiery cores of millions of exploding stars, and eventually galaxies appeared. Some galaxies exist in relative isolation, and others in large aggregations. Some galaxies are tranquil; others spew out energy with inconceivable violence. Quasars are at the fringes of the observable universe and generate a hundred times the energy of a typical galaxy. The Milky Way is powered by a powerful energy source at its heart, where radiation is emitted and fuelled by a black hole. Some galaxies are disc-shaped like the Milky Way and are termed spiral galaxies. Others are more spherical and are termed ellipticals. And still others have been violently twisted into asymmetry by the gravitational tug of galactic neighbours. There are also inexplicable fuzzy spots fixed in the heavens, known as nebulae, and the most energetic members of a whole group of peculiar celestial objects with bright nuclei are called quasars. Galaxies do not live in isolation and interact with fellow star systems, leading to near-misses and ruinous collisions. More severe interactions may cause their orbits to shrink and the galaxies themselves to combine. [4] It is estimated that about 99 percent of the universe, which is about 13 billion years old, consists of dark matter that is invisible to us. [5] The current theories about the ultimate fate of the universe paint a gruesome picture. For example, one theory states that the universe will continue expanding indefinitely, resulting in the "Big Freeze," but alternative models have also received much attention in recent years. In the Big Freeze scenario, the universe continues expanding, asymptotically approaching absolute zero temperature, and stars are expected to form for billions of years until the supply of gas needed for star formation is exhausted. However, existing stars will run out of fuel and stop shining, and the universe will slowly grow darker. Black holes will eventually dominate the universe, and they will fade away as they emit radiation. All material objects in the universe will disintegrate into bound elementary particles and radiation as the space between the galaxies increases, and eventually the universe will disappear. The beauty of the universe shines in the duality of light and darkness, in the birth and death of heavenly bodies, and with time, light will dissipate, the duality will end, and the universe will turn into a cold, lifeless darkness. [6] The observable universe is truly massive in size and majestic in its organisation, but it is not perfect and is temporal. All the celestial bodies will eventually collapse and disappear into oblivion, and an empty void will appear. There is no happy ending for the universe.

The planetary systems and the earth – Before the planetary system and the earth were formed, some essential chemical elements had to be present in the universe. Soon after the Big Bang, all the observable matter in the universe was compressed into dense plasma at extremely high temperatures. Then, as the universe cooled down, nuclear reactions took place between the atomic particles to bind them together to form the nuclei of light elements. Inside the stars, pressure increased dramatically, converting hydrogen into helium through a process known as "fusion." All the heavier nuclei like iron, carbon, nitrogen, and oxygen were made in the intense cores of stars, which exploded, spewing their products into the cosmos. [5] These products coalesced, growing steadily denser, and the heavier parts began to collapse under their own gravity, forming stellar bodies and galaxies throughout the universe. It is estimated that about 200 million stars had to explode so that we could be born. [5] Gravity caused the material to collapse into the solar system, creating a star and a disc of material from which the planets were formed. [7] The planets began as dust in orbit around the central protostar, which is a very young star that gains mass from its parent molecular cloud. The dust grains formed clumps, which in turn collided to form large bodies, and the large bodies increased in size through further collisions. Planetesimals are small planets that join under gravitation to form a planet.

The planetesimals are formed by compounds with high melting points, such as iron, aluminium, nickel, and rocky silicate. The rocky bodies became the terrestrial planets, and subsequent collisions and mergers between these planet-sized bodies helped the terrestrial planets grow to their present sizes. The large planets gradually migrated to new orbits, creating the planetary system we know today. The sun at the core of our solar system was formed by a similar process. [7] The sun was a ball of hydrogen and helium that was powered by fusion, and the temperature and pressure of the material inside increased, kick-starting the fusion of hydrogen atoms that helps the sun emit the life-giving light and heat to sustain life on earth. [8] The cosmos, stars, and galaxies were formed after a violent and explosive event at the beginning of time. The heavier elements of which the planets are made were created at the cores of the collapsing stars at extremely high temperatures and pressure. The visible universe's beauty is the result of the most violent and destructive events that could be imagined. It is interesting that God had no direct role to play in these processes, and the cosmos and the planetary system emerged of their own accord. The planet Earth was a barren and lifeless rock spinning around itself and orbiting the sun until life began.

The origin of plants and animals - The origin of life on earth has been the subject of much speculation, scientific research, and debate. It is agreed that for life to have started, carbon-based molecules, liquid water, and an energy source were essential. These elements could well have come from collisions of near-Earth objects, comets, and asteroids with the Earth, which caused the necessary biologic and geologic changes required to kick-start life. In the early stages, the Earth was bombarded by comet and asteroid impacts, which made the earth's surface too hot and prevented enough quantities of water and carbon-based molecules from surviving. This period is called the late heavy bombardment. Sometime after this period elapsed, in a short window, biological life began. When there was enough liquid water and carbon-based molecules on the Earth's surface, the building blocks of life appeared. Primitive life emerged from humble but very violent cosmic events, planetary mayhem, and harsh local environments. [9]

Once the basic elements such as air and liquid water were present on Earth, plants evolved to produce the wide range of species we know today. The evolution of plants started as early as 1 billion years ago, and more complex organisms emerged on land around 850 million years ago. It began with bacteria and single-celled organisms and progressed to freshwater green algae, land plants, terrestrial non-vascular and vascular land plants, and today's complex seed-producing and flowering plants. The vascular plants had vessels that could conduct and circulate liquids. [10] All animals and plants are made of a huge number of cells and owe their origins to a process where one cell ingests another but fails to digest it. Cells contain a nucleus and many internal structures, each surrounded by a membrane, that perform specific functions. Plants and animals evolved in the sea for nearly 600 million years until atmospheric oxygen levels rose to high enough levels to form the ozone layer, which protected living beings from harmful sun radiation and allowed them to move onto land. The first animals to appear on earth were soft-bodied, multicellular animals with a flat, quilted appearance. They were of different varieties, emerging from a lengthy period of evolution. These animals disappeared almost 544 million years ago and were replaced abruptly with other animals such as sponges, jellyfish, corals, flatworms, insects, earth forms, and leeches, which exploded in numbers over a period of 30 million years. The animals started to colonise the land about 530 million years ago.

Invertebrates (animals without backbones) and jawless fish (vertebrates with skeletal features) appeared about 505-440 million years ago, and eventually sharks and their relatives became more common in the oceans. As fish continued to evolve, groups of plants and animals colonised the land for the first time. Spiders, centipedes (elongated, segmented creatures with one pair of legs per body segment), and mites were among the earliest land animals. Four-legged land-based amphibian vertebrates, joint-legged invertebrate animals such as spiders, scorpions, ticks, and mites, and wingless insects then appeared. [11] Reptiles appeared on earth about 320 million years ago, and animals like birds, dinosaurs, and lizards evolved from them. [12]

Early reptiles moved away from waterside habitats and colonised dry land. This enabled them to produce eggs in which the embryo developed inside a membrane. Mammals evolved from reptiles about 286-248 million years ago. Some mammal-like reptiles gave rise to new creatures during the Triassic period, and the dinosaurs diversified into the dominant vertebrates. In the sea, ray-finned fish became the most dominant of all vertebrates. The dinosaurs and their ancestors dominated the land during the Triassic period, but mammals continued to evolve during this time. [11] Mammals have milk-producing mammary glands for feeding their young, and they became a diverse and rich group of organisms 125 million years ago. They include rodents, bats, hedgehogs, moles, shrews, humans, monkeys, lemurs, pigs, camels, whales, cats, dogs, and seals. [13] During the Jurassic period (213-145 million years ago), reptiles dominated the land, the sea, and the air. The dinosaurs ruled the land and the air in significant numbers but declined in diversity between 145 and 65 million years ago. There is also some fossil evidence to suggest a strong transitional form between birds and reptiles, though not all birds achieved powdered flight. Insects flourished about 145 years ago because of flowering plants, but a mass extinction 65 million years ago wiped out dinosaurs and every other land animal heavier than 25 kg, paving the way for the spread of mammals on land. At this time, fish became dominant in the sea. The mammals started to diversify about 65-55.5 million years ago and occupied many of the ecological areas. The primates appeared about 60 million years ago, according to fossil records, probably living in tropical or subtropical forests. Modern hoofed animals such as cows, pigs, and horses appeared about 55.5 and 33.7 million years ago, mainly in North America and Europe. As mammals evolved on land, they returned to the sea to evolve into animals such as whales. As the climate cooled between 33.7 and 23.8 million years ago, grasses and vast grasslands appeared. These changes helped further the evolution of browsing animals with teeth, such as horses. As land bridges formed, migrations of plants and animals took place. [11]

Mass extinctions have been a predominant event in the history of life. Extinction events occur when environments change, such as due to changes in glaciers, genetic mutations that result in the emergence of new species from older ones, organisms competing for habitat, or continental drift. The first mass extinction occurred 440 million years ago and killed a staggering 86% of all species. This was followed by more extinction events, the last one occurring 66 million years ago and destroying 76% of all life on earth. It is believed the worst extinction event happened 251 million years ago, destroying 96% of all living beings. [14]

The plants and animals that have populated the earth today have survived competition between varieties of the same species and between different species, disease, environmental and cosmic catastrophes and collapses, and unimaginable suffering and death. There is no evidence that any force other than these processes and events played a role in this difficult, and at times impossible, evolutionary path. Life, in all of its forms and varieties, evolved on its own to take on its final form. It seems that negative cyclic events in duality, such as birth and extinction, were the driving force behind the emergence of life on Earth and the wonderful natural world with all its complexities and varieties. There is no room for a benevolent creator like God.

The origin of humankind

The ape and primitive humans – Humanity evolved from different kinds of human ancestors, and there were coexisting species of human ancestors throughout our evolutionary past. The first hominids appeared almost 4 million years ago and disappeared about 1 million years ago. [15] Hominids are in a family that includes humans and some of the great apes. [16] Homo evolved from the hominid ancestors about 2.5 million years ago and eventually replaced them after coexisting with them for some time. Because it could make stone tolls, the first humanised hominid was named Homo habilis. Stone tolls enabled Habilis to cut meat into smaller pieces for chewing and crack open large bones to extract the fat-rich bone marrow. The tools were used to gather plants and small animals, dig, process, and share food. [15,17]

The anthropological human – The gathering of food was highly important to females with dependent offspring. As both male and female individuals participated in gathering food, they learned to interact together and share food for the first time, which was a unique achievement. In effect, the sharing of food provided a setting for socialising with others. Homo Habilis went extinct after 500,000 years. Homo erectus, the primitive humans, lived between 1.3 million and 300,000 years ago. They were toolmakers and hunters with the ability to control fire, and they possibly had some powers of speech. They travelled and explored the tropical regions of Africa, Asia, and Europe. Homo erectus survived for some 250,000 years after it became extinct. By this time, the seed of a family unit made of individuals who could hunt, interact together, cook with fire and share food, take care of their young, and communicate with the power of speech was in place. This was the beginning of structured human society as we know it today. The Neanderthals were cave dwellers who appeared almost 230,000 years ago and went extinct about 30,000 years ago. They were replaced by Homo sapiens (the modern man) 40,000 years ago but continued living side-by-side with the modern man and might even have contributed to the genes of the modern man. [15]

The religious human - As humans evolved and social gatherings and structures advanced, religions and religious belief systems took hold. There is no consensus among scholars as to when this transformation occurred, but it is agreed that it has had and continues to have a profound influence on human societies even today. It could have begun with the use of magic and progressed to religion, or it could have started with a belief in and practice towards unseen spirits.

Perhaps a brief review of some of the world's major religions will shed light on this fascinating and highly multifaceted subject. The religions of South Asia, China, the Pacific, the Americas, Persia, and Central Asia will be of interest, but due to their complexities, this review will be limited to some aspects of these religions.

South Asia – In the Indian religion, self-training, worship, and sacrifice are important. The first forms the basis of an inward search and a training of the mind. It is meant to discover the true self or attain liberation. Worship is directed towards a multitude of gods, and sacrifices of all sorts are confined to mostly plant life and melted butter. In later years, ritual sacrifices in the houses of gods or temples were dominant. Pilgrimages and festivals are also common, and God appears in various forms; different gods were manifestations of the same divine being. The theme of rebirth, or incarnation, is widespread. Each individual living being goes through many lives, and the person's destiny is shaped by the deeds she has performed before and in this life. The individual is bound to the round of existence, and when she eventually leaves it or at least overcomes it by great efforts, she will gain superior insight into the nature of the world. God is also involved in this process, and the higher side of the divine has a correspondence with the inner side of the soul. So, when an individual turn inward and practices meditation, he or she may experience that inner eternal self and, at the same time, achieve unity with the one divine being. [18]

China – The Chinese religion is very resourceful and colourful. The cult of the ancestors includes offerings to the ancestors as well as funerary rites designed to safely transport the dead to their final resting places. It is also worth noting the practice of balancing human activity with the spirit of the landscape. The goal of experiencing harmony with or identity with nature through following the ways of nature is highly encouraged. The concept of heaven was treated abstractly as providence and sometimes more personally as god. Rulers receive or do not receive the mandate of Heaven, which vindicates their rule. Many ethics revolve around family relationships, but monotheism provides an escape from the structure of the family. There are teachings about how nature works, rules of decency and good behaviour, meditation, loving benevolence towards other humans, and an acceptance that heaven or a personalised god resides above, and emperors follow his cult. [18]

The Pacific – In the Polynesian religion, there is a distinction between the invisible world and the world that can be seen and experienced, and both are interwoven. The gods are forces that are behind visible events, and they frequently visit human habitations. There is also a clear distinction between the heavenly world, where the gods reside when they are not with us, and the underworld, where the spirits of the dead pass. The gods are expected to visit where and when people want them to, and to achieve this, dwellings are prepared and incantations are formed. Once the gods arrive and people please them, they are sent back to where they came from. The gods are considered dangerous and must be treated with care. Statutes resembling the gods were made to entice them to take up temporary residence among the people. The gods were appeased by human sacrifices and incantations. Interestingly, there are various accounts of the creation of the world. One account says that god existed in the immensity where there was nothing. Existing alone, he became the universe, and all things in it, including wisdom, were made. [18]

The Americas – There were many empires in the Americas, but our understanding of their religions is incomplete because of a lack of written language. Temples were used for human sacrifices, and there were complex rites related to calendars describing the movement of heavenly bodies. The sun and moon were important deities, and there were images of gods in the form of animals. The emperor was responsible for the welfare of the empire and was the medium between his subjects and the great gods. He presided over earthly matters and was himself a god, the offspring of the sun. There were other mystical beings beneath the gods. In some cultures, the sun god became the focus of religion, and the cosmos was arranged vertically in seven layers of heaven, below which were the five layers of the underworld where the dead resided. In some other cultures, the universe was unstable, and the continuation of order required continuous sacrifices. The task of maintaining the order of the empire and the whole cosmos fell on the emperor. There was always a fear that the sun might not rise tomorrow, and the onward harmony of the world was in permanent danger. Since the cosmic and human worlds were in deep resonance, the sun needed blood, hence animal and human blood sacrifices. The universe was animated and controlled by various gods and spiritual beings, beginning with the supreme creator and a mix of male and female beings who fused oppositions in their own person. [18]

Persia and Central Asia – The Persian religions have had the most influence on the Abrahamic religions of Judaism, Christianity, and Islam. Monotheism is the main component of these religions. It is alleged that the Wise Lord fathered twin spirits. They each make a primordial choice, one for goodness, and the one for lie - a choice between good and evil. Every person faces the same choice. Good thought, the best truth, desirable power, great devotion, wholeness, and immortality are encouraged. There was opposition to blood sacrifices and the use of hallucinogenic drugs. God is perfectly good, and evil comes from him, is dependent on him, and yet independent of him. The adversary is evil because of a wrong choice. Good and evil struggle, and humanity must choose the good until a final stage when good and evil will be separated and the good will be rewarded with immortality after facing judgement. In the future, humans will have pure bodies in a state of resurrection. At the end of time, existence will be refreshed and made fabulous. The lie will end, the evil hostile spirits will be obliterated, and humans will be made immortal. [18]

- The South Asian religions shaped ideas such as incarnation or rebirth, liberation from the cycle of existence, a personal God who creates and recreates the cosmos as well as destroys and redestroys it, and an eternal something within the migrating individual with which the higher side of the divine has a correspondence.

- The core of the Chinese religion is appreciation for nature, good behaviour, a loving attitude towards others, and belief in a personal god who lives above.

- In Polynesian religion, there is a clear distinction between the invisible and visible worlds and the heavenly world and underworld. Gods visit humans when they are wanted, and then they are sent back to their abodes in the heavenly world. In the beginning, there was only god, and everything emanated from him.

- Religions in the Americas had strong elements of blood sacrifices in temples to bring stability and the continuation of order to the universe. The sun and moon were important deities, and there were also images of gods in the form of animals. The cosmos was highly structured but intrinsically unstable.

- The Persian religions were based on monotheism and good moral conduct. The struggle between good and evil will end, and if humans choose good, they will gain mortality and be given pure bodies. The world coming after the judgement will be free from evil and marvellous.

The emergence of religions and religious belief systems transformed the cultural, social, and moral landscape of humanity in a profound way. It was during this period that humans learned about gods, incarnation, rebirth, the soul, heaven, the underworld, and god's relationship with the visible world. The most influential ideas were about appreciating nature, good moral conduct, love for others, and monotheism. These concepts played a pivotal role in the formation and development of the Abrahamic religions. This most exciting and active period in human evolution laid the groundwork for the emergence of the spiritual human. However, humans' natural inclination towards paganism and obsession to appease gods by sacrificing human and animal blood in temples dedicated to them was the legacy of this period and was highly detrimental when God tried to raise a monotheist and law-abiding nation to serve Him.

The spiritual man – Spirituality is the greatest achievement of humanity. It is as if religiosity had to come first before spirituality could take root in the human soul and heart. The goal of spirituality is to emancipate humans from the destructive clutches of the duality of good and evil, which results in eternal death. Spirituality provides a channel for humanity's flaws to be addressed through faith in God. But first, the inner heart must long for the invisible God on its own merits.

There was a time when degeneracy was rampant among the children of men. Idolatry and immorality corrupted the affairs of men everywhere. At this time, the noble heart of humanity struggled to free itself from duality when a man yearned for the invisible God in his inner heart and rejected the ungodliness of his day. Abraham prayed and declared God Most High to be his God. [2] At last, after 4 million years of evolution, the godliness in the inner heart of humanity blossoms like a shining star in that darkness. Abraham was the seed of righteousness, and God was pleased with him and promised him descendants. (Genesis 17:1-2) God made a covenant with the descendants of Abraham, known as the Israelites. He gave them the ten commandments, or moral law, to make them a model nation worthy of serving Him to redeem the gentiles. Redemption was meant to transform the gentile nations to look more like the Israelites: God-fearing and law-abiding. The covenant was based on strict monotheism and commanded full obedience to the moral law. However, the Israelites struggled with keeping the moral law and often reverted to idolatry and immorality, displeasing God. It seemed that having Abrahamic faith was not enough to keep the people of God law-abiding, and a fundamental change was overdue.

When Jesus was born of a virgin, God wrote the moral law on his heart, creating a perfect being free from the clutches of duality. God was with Jesus during his conception, birth, life, and ministry, but not when he was on the cross, suffering death. When Jesus was on the cross, he cried, "My God, My God, why did you forsake me?" (Matt 27:46) God could not have interfered with Jesus on the cross because Jesus had to overcome death on his own merit. Light and darkness, life and death, right and left, are mutually dependent and can never be separated. Both aspects of duality—good, bad, life, and death—will be dissolved back to their original source, but what is noble and superior will survive the dissolution. [1] Jesus survived dissolution on the cross and passed through the final barrier, death, on his own merit because he was a perfect being with no deficiency in him that could have exposed him to closure by duality. All humans die because we have the duality of good and bad, hate and love, compassion and cruelty, faith and disbelief, and honesty and dishonesty in us. At last, humans can be changed into perfect beings in the person of Christ (redeemed), and with no duality in them, they pass through death, be resurrected, and escape the dissolution of duality like Jesus has done and inherit eternal life. Jesus said, "I am in the Father, and the Father is in me." (John 14:11) As mentioned already, God has duality in Him and must be in Christ to be made perfect and escape the dissolution of duality. Both God and humanity have found redemption in the perfect oneness of Christ. This makes the resurrection of Christ the most important event in the last 14 billion years since the universe was made and redemption the only escape route from the dissolution of the world, when all things, including light (God), will disappear into darkness, where it all came from. [6] Jesus claimed to be the life and resurrection (John 11:25), and outside of him, all things will perish. The Bible teaches that death precedes resurrection, but this is not true. Because Jesus was a perfect unity free of duality, he could never have died. Jesus went through death victoriously because his resurrection was assured by his perfect unity beforehand. It was for the sake of the elect that he bore the agony of death on the cross. In duality, death is the end of everything. (Philip 19) [1]

The main goal of creating the universe was accomplished after Jesus triumphed over death on the cross and was resurrected. God redeemed Himself in Christ and offered the same to humankind. The next stage was to preach the good news, bring this present age to an end, and usher in the Kingdom of Light for the redeemed.

Summary – The most important event in this unique period of human history was the emergence of faith in the invisible God in the inner hearts of men. What was noble in human nature naturally freed itself from the clutches of duality. God then blessed the inner faith with the moral law to make it perfect unity in the person of Jesus Christ. In Christ, both God and man are redeemed. Whoever receives purification in Jesus Christ will escape the horror of returning to darkness when duality is dissolved.

In the next chapter, God's providence at the cross of Christ and the mystery of the Christian cross to put an end to the wicked men's scheme will be discussed.

Chapter 2

An empire ending all empires – End of history

When the history of humanity is examined retrospectively, a disturbing pattern emerges. Throughout history, empires have risen to bring glory, power, and wealth to a few and then collapsed into ruins. Their fall has caused so much suffering, insecurity, death, and displacement for their subjects. This pattern of human existence could have continued forever unless God intervened to bring it to an end. God Christianised and revived the pagan Roman Empire into the Papacy and used its resources and military might to spread the good news of the gospel of salvation to the four corners of the globe. However, the legacy of invasion, conquest, and acquisition of land, labour, and resources remains at the forefront of Western foreign policy and aggression against other nations to this day. This is an unwelcome legacy from the Christian past that has caused so much bloodshed and suffering.

When the Cross became a sword – The destiny of humankind has been interwoven with the rise and fall of empires. Over the last four thousand years, there have been over 200 empires. Some empires lasted 90 years, while others lasted 2334 years. [19] The Roman Empire was one of the greatest empires to ever appear on the European continent. Rome was a republic for 450 years and became an empire in 27 BC under the rule of the first Roman emperor. From the early days, Rome won numerous military victories, and literature, religion, and art flourished. The strong, well-equipped, and well-trained military was the key to Rome's success in controlling Asia, Africa, and most of Europe. At the end, the empire was divided into the Eastern and Western Roman empires. After years of military conquest and expansion and 700 years of rule, the Western Roman Empire started its slow collapse, and by 476 AD, it broke into small kingdoms and fell into the Dark Ages. [20] The Eastern Roman Empire, called the Byzantine Empire, continued to exist until 1453, when it was conquered by Ottoman forces. [21] The Romans advanced in military technology and planning, architecture, building, road planning and construction, public health and welfare, the Roman alphabet and Latin as an international language, law and justice, religion, customs, clothing, ethical issues, technology, engineering, trade, and games. [22, 23] The Roman Empire will be remembered for its immense appetite for invasion, conquest, acquisition of properties, and subjugation of people to its rule. Slaves frequently rebelled against their Roman rulers, but rebellions were suppressed brutally by the Roman army.

In 312 AD, one hundred and sixty-four years before the collapse of Rome, the Roman emperors Constantine and Maxentius were in battle on the Milvian Bridge on the Tiber. It is alleged that Constantine and his soldiers had a vision sent by the Christian God, promising victory if the first two letters of Christ's name were painted on the soldiers' shields. [24] Other historians claimed that Constantine had a vision, saw a shining cross, and inscribed upon it an instruction that said, "By this sign, you will conquer." Another account explains that Christ himself appeared with the cross, and the order to be victorious was sung by angels. Jesus said he did not come to bring peace on earth, but a sword. (Matt 10:34) He also said he had cast fire on the world and was watching over it until it blazed. [1] The event at the Milvian bridge fulfilled Jesus's prophecy.

Constantine formally converted to Christianity in 312 AD, and in 313 AD, Christianity was decriminalised. Constantine organised and attended many church councils. Bishops from all corners of the Roman world attended to thrash out a doctrine for the Church. In one meeting, Constantine dressed in his bright purple robe, embroidered with gold and inlaid with stones, and then sat on a small golden chair. [25] It is interesting that the Popes of the Catholic Church follow the tradition set for them by the Roman emperor rather than by Jesus Christ, who wore sackcloth and sandals. The meetings were strictly monitored by the emperor, who took an active and forceful role because the bishops could not agree among themselves. [23,25] The exact intentions of Constantine were not obvious, but it is highly unlikely he had the best interests of Christ or the Gospel in mind.

It was in 380 AD that the state church of the Roman empire was formally declared. State and church became one body, and the fate of the Christian gospel and the continuity of the Roman Empire in the form of the Papacy intertwined forever. [25] It could be argued that God Christianised the Roman empire and then revived it through the Papacy after its collapse to spread the gospel of salvation to the four corners of the world. On the Milvian Bridge, the cross was placed in the hands of a Roman Caesar to become a sword of conquest and conversion, and he was given unconditional and unlimited power to conquer in Jesus' name. The heathens could only be converted to

Christianity by the power of the sword, and this was the Roman cross on which Jesus was crucified. Jesus claimed ownership of the might of the Roman empire through his crucifixion on the cross to spread the good news to the world. The Christian sword showed no mercy to anyone and set the course of human history on its final leg.

The history of the Catholic Church and the Papacy is marred by violence and bloodshed on a truly horrific scale. The Catholic Church waged war against those whom it condemned as heretics and apostates. The victims were mainly Jews, but there were also religious wars with Muslims, referred to as the Crusades. There were nine Crusades for the conquest of the Holy Land, starting in 1096 AD and ending in 1291 AD. Other crusades took place between 1209 and 1320 AD to eliminate heretics and rescue puppet kings. Inquisitions were waged from 1184 AD to 1798 AD against people and movements considered to be apostates or heretics to Christianity, including Jews, Jewish converts to Catholicism, Protestants, printed literature, witchcraft, Hindus, Muslims, and superstitious practices. The inquisition covered huge geographical areas, from Europe to North, Central, and South America, Asia, Oceania, and Malta. There were also blood libel and desecration accusations against Jews and frequent expulsions of Jews from their homes. [26] The Catholic Church lost its power and independence in 1798 AD when the residing Pope was removed from power, but in 1801 AD, some of its civil status was restored. [27] In 1929, the Vatican territory became a sovereign country. The Vatican City State is now governed by the Holy See and the Pope, who is the bishop of Rome. [28] The Papacy might have been responsible for over 100 million deaths to date. [29] There is no denying that the Catholic Church has been instrumental in spreading the gospel of Christ to the heathen in the four corners of the world and enriching the cultures of many people. The Catholic saints have been great examples of the Christian faith and good conduct. But as history shows, the Papacy has a dual nature, good and bad, which is common to all things (Genesis 1-5). The church has apologised for its past mistakes and is now showing a more compassionate and constructive face to the world. The Church of Rome faced many serious challenges to its authority and existence in the past, but the Protestant Reformation in the sixteenth century was by far the greatest of all.

The Protestant Reformation and the challenge to the Papacy – The Protestant Reformation started in 1517 in Germany and was triggered by the rejection of the practices of the Catholic Church, which were considered corrupt and worldly. It was based on two premises: we should rely on Scripture alone, and our salvation depends not at all on work but on faith alone. The idea that the Church alone had knowledge and understanding of the Bible was challenged and then rejected. After the Bible was translated and the printing press was invented, the movement spread, piety, love of God, and salvation that can only come from God took root, and dissatisfaction with the hierarchy of the Church grew. As the Reformation progressed, some advocated for the separation of religion and politics. One idea suggested that if salvation is by God alone, then he has already chosen who is to be saved and who is not. What a person does has no bearing on his or her divine destiny. In other words, individuals are predestined for salvation. The preaching of the Word became highly important, and major alterations were made to the plans and layouts of the churches to make preaching easier and more inclusive. [18] The Reformation divided Europe into Catholic and Protestant camps, and this was followed by thirty years of war in the 17th century, though the conflict continued until the early 18th century. This period is referred to as the "European Wars of Religion." [30] The Papacy's power, prestige, influence, and wealth diminished greatly. The Protestant Reformation elevated reason above authority and blind faith, both of which were major challenges to the Church's dogmas. The Papacy reformed itself but kept some structures that the Protestants rejected in place. The fathers of the Protestant Reformation identified the Pope as an antichrist and his Church as a harlot. This posed a major challenge to the Papacy, and to counter the Reformation and assert Catholic values, the Church created the Society of Jesus, or the Jesuit order. The Jesuits received military and highly intellectual training, as well as spiritual exercises for serving Christ in the world. One effective tool that the Church created was a missionary department that led to modern-day "propaganda." [18]

Maritime expansion, colonisation, and enlightenment – After the Roman Empire collapsed in 476 AD, Western Europe fell into the Dark Ages. From the 15th to the 17th centuries, the age of discovery began. Seafaring Europeans explored, conquered, and colonised vast areas across the globe, starting with the Americas. As time passed, many states adopted colonialism as their foreign policy. The colonisation steamed ahead with vigour and ruthlessness, covering the Mediterranean, Africa, India, Asia, the Pacific Oceans, Australia, and the Atlantic archipelago. The polar regions were explored later. The growth and expansion of international trade brought unprecedented wealth and influence to European nations. At the same time, new diseases brought in by the Europeans destroyed millions of people, with the Americas and the Pacific regions suffering most. Countless numbers of native populations were killed by widespread enslavement, exploitation, military invasion, and forced conversion to Christianity at the edge of the sword. As European economic influence and trade grew, the Europeans started to dominate the world's affairs. [31] The immense wealth and improved lifestyle set the scene for the next chapter in European history, the "Western European Enlightenment."

The Western European Enlightenment in the 17th and 18th centuries paved the way for major advances in Europe. A more scientific approach to history was gaining traction, and there was much debate over the truth of scripture and the authority of the Catholic Church, further undermining its authority. The foundations of capitalist economics were laid in the early 18th and early 19th centuries. This was followed by scientific and industrial revolutions that led to the inventions of steam engines, the cotton industry, steam locomotives, electric dynamos, and ether, first used as an anaesthetic in an operation, which changed society in a way that was never expected. The use of crop rotation and various reaping machines in agriculture aided European colonisation of the North American steppes. The scientific and industrial revolutions transformed Western European countries into rich and powerful nations with formidable military forces. [18] The appetite for conquest and the acquisition of land, resources, and manpower grew even more, leading to major wars and immense destruction in both Europe and overseas.

In the 19th and 20th centuries, the European continent witnessed mechanistic wars the likes of which had never been seen before.

After the European empires were decimated, new independent states emerged, and the European continent that we know today came into existence. In the 20th century alone, millions of people lost their lives, mostly on the European continent. [32] The Western powers' aggression did not spare nations in Africa, South and Central America, the Middle East, Southeast Asia, and the Indian subcontinent. For example, in the recent wars in Iraq and Libya, over 1 million people were destroyed. The empire that began with the establishment of a small settlement on the Tigger River, grew into a republic and a powerful empire under the Caesars, and was then Christianised by Jesus Christ to spread the good news to humanity, has lasted nearly 2500 years, making it the longest and most powerful empire ever to live. It continues to unleash havoc on kingdoms and nations all over the world, even today. The empire did well for the gospel of Christ by converting countless heathens to the faith through the power of the sword. However, the West is secular in its outlook, and invasions and conquests are now conducted for profit at the expense of countless innocent lives and so much material destruction. The Western way of behaving is still shaped by its Christian past, when it was said, "Go and conquer in my name." This is a legacy that has done so much harm to so many human beings in post-Christendom Europe and will stay with us until the end of time.

Summary – The rise and fall of empires, which have been the hallmark of human presence on earth, will soon come to an end. The Christianisation of the pagan Roman Empire before its collapse and the revival of the Empire through the creation of the Papacy after its fall were turning points in history. The Protestant Reformation challenged the legitimacy of the Papacy and identified the Popes of the Catholic Church as antichrist. The Jesuit order of the Catholic Church devised an unimaginably effective and brilliant scheme to counter the Protestant Reformation and its identification of the Pope as the Antichrist and his Church as a harlot. This scheme is so sophisticated and well thought out that it is hidden from sight and mind, but it has been in full swing for some time. It is the greatest deception in modern history.

When Jewish history and its interactions and dealings with the Papacy and the Western European Enlightenment are examined side-by-side, an interesting picture emerges. The reason for the establishment of the state of Israel in the Middle East, its implications for world affairs, and the fulfilment of biblical prophecies will be examined in the following chapter, considering the Jesuits' Counter-Reformation.

Chapter 3

The great deception – Israel in the Holy Land

There are frequent warnings in the Bible guarding against deception in the end times. It is stated that even the elect will be deceived. The deception is religious, and to understand its nature, we need to look back at the Catholic Counter-Reformation. The main goal of the Counter-Reformation was to divert attention away from the Popes of the Catholic Church, who were identified as antichrist and their church as a harlot. This has been accomplished through an elaborate and well-planned scheme to have a temple constructed in Jerusalem, where a man of Jewish ancestry will supposedly appear and claim to be the Messiah. Jesus Christ never *sanctioned such prophecies, and there is no biblical justification for why the Jews should have ever returned to the Holy Land unless by deception. However, their return fulfilled their millennium-old inspiration for their ancestral land and the planned rebuilding of the Third Temple in Jerusalem. At the same time, it has provided the Papacy with an opportunity,* albeit accidental or deliberate, *to remove the challenge of the Protestant Reformation against its Popes and institutions once and for all.*

The end-times Prophecies in the Bible are complicated to understand, and any personal views stated here may be controversial. Every day, we are presented with different scenarios of how the world will end before the return of Jesus Christ. Some are based on the Bible, and others are purely conjectural and for the entertainment of the masses. Often, future events do not happen as we may imagine and can take an unexpected turn. Major institutions and structures will be destroyed in a global conflict. This will cause mass destruction, untold suffering, and horror, and no one can know what will emerge from the ashes of the coming conflagration; we can only speculate. This period is referred to as "tribulation." The scenario that the author presents here is one in which the Papacy will play a major role in end-times events. Under no circumstances do the author's views reflect on those of the Catholic faithful. Because of its dual nature (cross of salvation and sword of destruction), the forces of hostility and their principalities, guided by satanic inspiration and impulses, will target the institution of the Catholic Church, the only remaining body of the Christian faith and devotion in tribulation after all else has been destroyed, to unleash their agenda of destruction on humanity. This may include a complete takeover of the church and its resources. Historical events have shown that the Papacy indeed has a weakness due to its dual nature, which may be exploited by hostile spiritual forces to transform it into a demonic entity. Whatever the future holds, it is fair to assume that it will not be purely random events and processes but will have a decisive Judeo-Christian dimension to it. This will put the Catholic Church at the forefront of the coming end-time events. But first, the historical context in which this situation is taking place must be examined. The Protestant Reformation has had a profound impact on the Catholic Church and is playing a major role in developing the scenario being proposed.

The Jesuits' counter-reformation – The Jesuits' master plan was to restore the legitimacy of the Pope and the Papacy, which had been destroyed by the Protestant Reformation. For this plan to work, it was essential that a temple be constructed in Jerusalem, where a false Messianic figure would appear, then be destroyed and replaced by an allegedly true Messianic figure from the Catholic Church. This would restore the credibility of the Pope as a Christ figure and the Papacy as the true Church. The Jewish people were the only nation that had claims to the Holy Land, and temple worship was part of their religious tradition. The Jesuits' master plan and the longing of the Jews to return to their homeland in Palestine converged to set the scene for what will be coming to the world soon. But there had to be scriptural support for the Jewish return to Palestine, and some Bible verses have been used to this end. For example, God promised that he would take His people from the nations, gather them out of all the lands, and bring them into their land. (Ezekiel 36:24) This undertaking and many others like it were part of the old covenant. But Jesus Christ replaced the old covenant with a new one, and hence, the promises and undertakings under the old covenant were no longer credible and needed to be fulfilled. The selective use of Bible verses to support claims to a land that was promised to ancient Israel under strict conditions of a covenant that was nullified centuries ago is not legitimate.

Jewish return to the Holy Land - Although some Jews in diaspora longed to return to their homeland in Palestine because of severe persecutions, the majority did not wish to do so. The Western European Enlightenment promoted a secular view of the world, weakened the authority and influence of the Catholic Church, and made possible Jewish emancipation, freeing them of their age-long civil and political disabilities. [33] Most Jews in Europe were reasonably well assimilated into their societies, despite anti-Semitism, and were reluctant to leave the comfort and modernity of the West and immigrate to a backward and insecure land in Palestine. But the Holocaust changed all that. It destroyed half of European Jewry and forced Jews mostly from Europe, Russia, the Middle East, and some parts of Africa to immigrate to Palestine. What ancient religious inspiration and longing failed to achieve for centuries, an extreme act of violence achieved within a few years. It is worth remembering that some of the hierarchy of the ruling German elites were devout Catholics and visited the Vatican, and the Pope might have had full knowledge of the ongoing holocaust. [34] The Catholic Church's involvement in the Holocaust, if any, raises serious questions about its motivations, but it might have fit well into the Jesuits' master plan. It is no exaggeration to say that since the Jews were expelled from the Holy Land almost two thousand years ago, their fate, character, and destiny have been shaped by intense Christian persecution. Those who lived outside of Christendom did better, but those who remained under the rule of the Roman Church suffered horribly. [26] After suffering from pogroms, frequent expulsions, confiscation of properties, and inquisitions, the Zionist movement emerged in the 19th century to re-establish a Jewish national home in Palestine. The goal was to encourage the emigration of millions of Jews from the diaspora to Palestine and the establishment of a Jewish state. [35]

Scriptural legitimacy for the Jewish return to the Holy Land – When Jesus visited the Temple in Jerusalem, he prophesied its destruction. (Matt 24:1-2) The coming destruction of the Temple also implied the expulsion of the Jews from the Holy Land. This happened in 70 AD, when the Roman armies destroyed the Temple, killed, and deported the Jewish inhabitants of Judea and the surrounding areas into exile. Jesus did not mention the future return of the Jews to the Holy Land or the construction of a third temple in Jerusalem. The biblical narrative does not support the need for the construction of a third temple since the first and second temples had no benefit to God or ancient Israel and were subsequently destroyed by foreign armies. How can the construction of a third temple be of any benefit to God and humanity when the old covenant was replaced by the new one centuries ago? Moreover, since the ascension of Jesus Christ, the body of Christ, where redemption is, has been the abode of the true Israel and should be the focus of the faithful's devotion and hope. What does a man gain by putting his trust, hopes, and inspirations in man-made structures, meaningless rituals, and worthless theatrics? Jesus was highly dismissive of the religious leaders and practices of his time, and when he visited the Temple in Jerusalem, he called it the den of robbers. (Matt 21:12-13) It has been widely accepted that the Jewish people must have a homeland, but there is no biblical reason why that should be in Palestine. There are anti-Zionist factions among Jews who reject the State of Israel. For example, Orthodox Judaism strongly opposes Zionism because it espouses nationalism in a secular fashion and uses "Zion" in literal rather than sacred terms to achieve its goals in this world. [36]

The Balfour Declaration – After the Zionist movement was born, the Balfour Declaration of 1917 paved the way for the return of the Jews to Palestine. [37] The Declaration stated its support for the creation of a home for the Jewish people in Palestine, which was under the British Mandate at the time. Both the Zionists and the anti-Zionist

Jews participated in the negotiations, but there was no representation from the local population in Palestine. The Palestinian Arabs comprised the vast majority of the local population, but their views were not taken into account, and there was little or no protection for their rights. As a result, it led to the ongoing Israeli-Palestinian conflict, which has been raging for years. The state of Israel was formally announced in 1948. [38] The Roman Catholic Church abolished anti-Semitism and the persecution of Jews in 1964 and established relations of mutual understanding and respect for the Jewish people. Jews were no longer considered rejected, cursed, or guilty of killing God. [39] The Jewish cause gained substantial support from the Zionist Christians, who have been and continue to be the main backers of the State of Israel and the rebuilding of a temple in Jerusalem. [40] It is a great mystery that, after over a thousand years of persecution, the Papacy is supporting Judaism and the Jewish State almost overnight. This engagement will define itself in a more profound way when the Third Temple is built in Jerusalem.

The Temple Institute is preparing a blueprint for the construction of the Temple in Jerusalem. Sacred Temple vessels for worship, the High Priest's breastplate containing the 12 precious stones of Israel, musical instruments, garments, and priests in charge of sacrificial duties and caring for the Temple, its vessels, and furnishings are being prepared for the Temple. [41,42] There are disputes with the Muslims regarding the status of the Temple Mount, and until a consensus is reached, a date for the construction of the Temple will remain in the future. It will be interesting to see how this dispute will be resolved, but there is no doubt that a Jewish temple will rise in Jerusalem sooner or later.

The dynamic behind the construction of the Third Temple – The rebuilding of a temple in Jerusalem represents the meeting of two historical events: the Jewish longing to return to the Holy Land and resume temple-centred worship; and the Jesuits' plan to counter the challenge to the Pope's legitimacy and accusation of antichrist that was levelled at it by the Protestant Reformation. These events are converging to set the scene in the Holy Land for an apocalyptic end to human history. Whether this convergence of interests is accidental or deliberate is not clear, and we can only speculate on its consequences. The Bible mentions that a day will come when the man of sin, the one opposing and exalting himself over everything being called God or an object of worship, will appear. He will sit in the temple of God as God, showing himself that he is a god. (2Thessal: 2:1-5) It is plausible that Jesus is referring to the hearts of men as the temple where God sits and as the place of sin and abomination before God. We live in an age when people are ungodly, selfish, and arrogant at heart and act as if they are gods. There could also be a physical dimension, i.e., a temple in Jerusalem, where this prophecy will be fulfilled.

What could possibly happen when the Third Temple is built? Consider this scenario: The Temple is built, and a person of Jewish ancestry sits in the Temple and claims to be the promised Messiah. This will be achieved through a covenant with the Papacy, which will be the sole religious authority in the world. This is not far-fetched since the Papacy is uniting the different religions under its own banner. Recently, Pope Francis urged members of all religions and those belonging to no church to unite for peace, justice, and the environment. [43]

A Jewish Messiah, a military figure, can emerge as the nation's saviour during a period of major military conflict and turmoil that threatens the State of Israel's very existence. Sometime during his Messiahship, the covenant with the Papacy will dissolve, and the Jewish Messiah will be named the Antichrist and killed. Then, a figure from the Catholic Church's line of Popes will take the seat in the Temple and claim to be the true Messiah and God, fulfilling the prophecy in 2Thessalonians 2:1-5. In the end game, the Papacy and Judaism come together as a final act to bring human history to its foregone conclusion. The stage on which the human drama has been unfolding for millennia will dissolve into its beginning, and the Judeo-Christian narrative will set the scene for what is coming.

The second coming of Jesus will not occur until many Jews are converted to Christianity, according to some Christians. The missionaries believe that the required mass conversion of the Jews will not take place until the time of the great tribulation in Israel, when a significant number of the Jews will be killed and the remaining ones will convert to Christianity. [44] It remains to be seen how events will unfold in the coming years in the Holy Land. Whatever the outcome, there is no doubt that Israel will play a major role in the fulfilment of the end-time prophecies in the Bible.

The Antichrist will not rise until a system of government is in place to support his rule over all of humanity. The Bible refers to two entities that will rise: the beast and the dragon. The dragon will give its powers and authority to the beast. The dragon was worshipped for giving the beast authority, and the beast was worshipped. (Revelation 13: 1-7) The Bible teaches that all men will be given a mark on their right hand or on their foreheads. No one will be able to buy or sell unless they have the mark, the name of the beast, or the number of its name. (Revelation 13: 15-17) The mark on the right hand signifies actions, and the one on the forehead signifies thoughts.

How will such a system look, and how will it operate on a global scale? Consider this scenario: All currencies are replaced by a centrally issued digital currency. Every qualified person is allocated a fixed amount of the digital currency per month. All property rights and ownership are revoked, and no one owns anything. To receive the allocated digital currency, one must be chipped in the right hand so one's actions, such as buying and selling, can be monitored. Transactions can be carried out if individuals are loyal to the central authority. The thoughts of the chip-holders will be monitored by inserting a microchip device on the forehead and possibly by using artificial intelligence (AI) that will report the brain's functions to a central database. If a person is judged to be hostile or undesirable, the microchip can be turned off, and the person will no longer be able to buy or sell. The authority in question can well be the Papacy, collaborating with a political or government entity. Is this implausible?

The World Economic Forum (WEF) is the International Organisation for Public-Private Cooperation. It brings together political, business, cultural, and other leaders of society to shape global, regional, and industry agendas and is the most influential body running the world. In a press release from the WEF, one of the objectives is to achieve zero private ownership by 2030. [45] The elimination of private ownership and the imposition of digital currency will create a system of control unprecedented in human history, the "beast system," with the Papacy playing a central role in this process. [46] But the faithful who go to the Lord God will remain under his protection and escape the coming destruction by the beast and the wrath of God. (Psalms: 91: 1-4)

Summary – Christianisation resurrected the Pagan Roman Empire as Papal States, launching it on a course of invasion, conquest, and colonisation that is still being pursued by secular Western European countries and America today. The Protestant Reformation identified the Popes of the Catholic Church as Antichrist and the Papacy as the Harlot. The Jesuit order was created to counter the Reformation. The Jesuit master plan and the desire of the Jews in diaspora to return to their homeland in Palestine have converged, eventually leading to the building of a temple in Jerusalem. A messianic figure of Jewish ancestry will appear in the Temple through a covenant with the Catholic Church. This figure will be pronounced the "Antichrist" and replaced with one from the Catholic Church who will claim to be the true Messiah and God. In fact, he will be the man of sin, and Jesus will destroy him in the final act to bring the age to an end and usher in the Kingdom of God. There has never been a biblical reason why the Jews should have returned to Palestine, and Jesus never sanctioned their restoration to the Holy Land. Therefore, this is a man-made deception and has no legitimacy in the Bible. It is folly to underestimate the power of the Papacy behind the scenes. The sequence of events that have taken place since the birth of the Zionist movement, leading to the creation of the State of Israel, has the hallmarks of political and strategic planning by men in high places of power and influence rather than divine providence. Whatever happens in the future and however the coming events unfold, there will be a purposeful Judeo-Christian dimension to them, that for now, we do not understand and can only speculate on.

There is no doubt that the events taking place in the Middle East today will contribute to the fulfilment of the biblical prophecies of the end times, but salvation in Jesus Christ must take centre stage and remain the hope and inspiration of the faithful.

Chapter 4

Summary and conclusions

The best way to search for the truth is to expose deception. The truth in the Bible is wrapped up in stories, metaphors, symbols, parables, prophecies, and confusion. But when these layers are removed, the truth shines like pure light in the darkness of ignorance. The light shining in Jesus Christ is too bright for the blind at heart to see, and deception is against it. The root of this deception lies in the recent history of the Christian West.

The biblical creation view and evolution – Genesis 1–5 of the Bible tells the story of creation, when God separated light and darkness, resulting in duality. The duality is a medium in which everything came into being on its own over billions of years of evolution. The world that we know today, with all its beauty, complexities, and wonders, is the product of unimaginable violent and destructive processes that obliterated millions of living creatures. Jesus frequently referred to God as a gardener. (Matt 13: 410) Let's consider a gardener.

A gardener plants an apple seed in the soil. The soil provides an environment in which the seed can grow into a plant. The soil is made of organic and inorganic matter and is a source of water and nutrients. Processes such as leaching, weathering, and microbial activity combine to create a whole range of different soil types. Each type has strengths and weaknesses for agricultural production. [47] The soil's poor quality is due to poor drainage, the presence of blights and crop pathogens, a large population of predatory insects, a large population of weeds, and harmful bacteria, earthworms, and bees. [48] The soil's good quality is due to its good tilth, sufficient depth, proper level of nutrients, good drainage, large populations of organisms, and resistance to weeds and degradation. [49] Normally, a garden has a combination of all these features, and thus it provides a dual environment for the seed to grow in. The seed then produces a small tree, which grows to have branches and leaves, and eventually apples grow on the branches. While the tree is growing, it experiences hot and cold, light and darkness, wet and dry, wind and calm, and invasions by insects and birds. Some are beneficial to the tree, while others are detrimental. This is the dual process by which the tree grows to produce apples. The gardener has no role to play in the process other than pruning the tree branches from time to time to keep them healthy. He then collects the apples when they ripen. Is it right to say that the gardener created the apples? No. The gardener planted the seed, and the tree produced the apples. The gardener may choose to sow seeds to produce other fruits. For example, apple seed is made of fatty acids such as linoleic acid, palmitic, linolenic, stearic, and oleic acids; cherry seed is made of oil, proteins, and dietary fibres; and orange seed is rich in saturated and unsaturated fatty acids and bioactive compounds such as carotenoids and flavonoids. [50] These seeds are very different in composition, but when they are sown in soil, they grow into trees and produce different fruits. The seeds are in the soil, and the trees grow in the open air. Hence, it could be argued that duality (different conditions in the soil and the open air) produces these products. The gardener simply selects the seed, and the soil and fresh air do the rest.

Likewise, God did not create anything, but the duality of light and darkness, as stated in Genesis 1-5, provided a medium in which all things came into existence of their own accord over billions of years of evolution. God is the seed-maker par excellence and not the creator, as it is understood in the Bible. At the beginning, God separated light from darkness to form duality in which the universe appeared, and then He placed a seed of life in the earth, and the seed grew into a tree of life over billions of years, producing plants, animals, and human species. If God wills it, he may sow another seed in this duality to create another unimaginably exotic and marvellous universe. This is the genius of our Lord God. We do not know what other seeds God has in store, but we know that the soil (resembling a duality) is fertile and highly creative, though not perfect, and can produce precious things, such as righteous people like Abraham, who can be made perfect like Jesus Christ.

An empire to end all empires – Christianity revived the pagan Western Roman Empire into the Papacy and placed it on a trajectory to bring the age to an end. This period in human history saw a significant increase in the power and influence of the Papacy, as well as the spread of Christianity and the good news of the gospel of Christ among the heathen people. This was achieved mostly by invasion, conquest, forced conversion, and rigid rules. However, after the Western European Enlightenment, when modern secular states emerged in Europe, invasions, wars, and colonisation continued even more forcefully and became the Europeans' primary preoccupation and foreign policy. In this period, the Christianisation of the conquered people continued with the blessing of the Catholic Church, enriching its coffers and spreading its influence further afield. This is the legacy of the Christian era, which served the gospel of Christ well but brought so much unforgivable devastation, bloodshed, misery, and crime against humanity in the last 200 to 300 years. The Papacy and its satellite states have committed atrocities on a grand scale and are responsible for countless deaths. [26] It must be mentioned that the Catholic Church throughout its history has produced great saints, and most Catholics are devout believers. However, the Protestant Reformation questioned the legitimacy of the Popes and the Catholic Church's authority. The Counter-Reformation has devised a scheme to restore the credibility of the Papacy, and the Middle East is the epicentre of this grand scheme.

What should our devotion and hope be? – We are witnessing unfolding events in the Holy Land, with the end game being the building of a temple in Jerusalem. The main focus of our interest and devotion should be the gospel of Christ and the lordship of Jesus Christ as the cornerstone of our salvation. However, extensive and relentless media coverage of Israel, the building of a temple in Jerusalem, and all the work being done to prepare a priesthood for the running of the temple have diverted our attention elsewhere. Behind this campaign lies a master plan to present the Pope of the Catholic Church as the ultimate source of hope and salvation for humankind, the Messiah. There is no doubt that when the temple is built and the Messianic figure appears, nations will be mesmerised by the events and theatrics unfolding there. There will be miracles, blood sacrifices, musicals and festivities, healing, and revelations. [41] This will be an important component of the beast system, assisting in the deception of people. Christians alive in Christ are seriously in danger. Those who are not in Christ will not die because they never lived so as to die. One who has believed in the truth has become alive, and there is a risk that this person may die because of being alive. (Philip 3) [1] We are reminded that we live during deception, and in their deception, they refuse to acknowledge God. (Jeremiah: 9-6) The words they utter are wicked and deceitful, and they fail to act wisely or do good. (Psalms: 36:3) We are commanded to put on the armour of God to be able to stand against the wiles of the devil. Our battle is not with flesh and blood but with rulers, authorities, the darkness of this age, and the spiritual power of evil in the heavenlies. (Ephes: 6:10-12) It is in our interest to remember that salvation is in Jesus Christ alone, and no source outside of him, no matter how impressive and influential its outfit, scope, teaching, or resources, will ever save us from what is coming. There is no salvation in the world of duality; it only comes from the perfect oneness or unity of Jesus Christ.

Summary

- We are witnessing events that will lead to the building of a temple in Jerusalem and the rise of a new world order that will exert total control over the affairs of humankind. This new world order is Satanic by nature and has a strong eschatological element to it. It is the culmination of Jewish longing to return to the Holy Land and centuries of planning by the Catholic Church's Jesuit order to counter the Protestant Reformation's accusations levelled at its Popes and institutions. It is in denial of the lordship of Jesus Christ and rejects his authority.

- When Jesus Christ passed through the barrier of death victoriously, was transfigured into eternal light, and ascended into heaven, the purpose for which the universe was created was fulfilled. Now, all good things will be made perfect in Christ, and we will escape the clutches of duality and its dissolution. The history of the world begins with the duality of light and darkness (alpha) in Genesis and ends with the perfect oneness of Jesus Christ (omega) in Revelation.

- Will the righteous ever die? No. When a faithful person dies, by God's grace, the person will be redeemed in Christ, be transfigured into a being of light, and be made a citizen in the Kingdom of God, as Jesus did after he went through death on the cross. Therefore, for a person saved in Jesus Christ, death is a transition to the kingdom of God and not a termination. For others, death after judgement is closure for all eternity.

- All things have begun in duality, where there is light and darkness, right and wrong, good and bad, and love and hate. This duality has shaped the good and the bad in their final forms through an unimaginable struggle since the dawn of time. In humility and compassion, the author of all that is pure and noble made good flesh in Jesus Christ and revealed it to the righteous. The bad has its fate built into the fabric of duality, which will dissolve back to its original source at the end of the age.

Death nullifies humans, and only God's grace can resurrect us in Christ. Death did not nullify Jesus, who was resurrected through his perfect unity before he went on the cross. Since death has no intrinsic element of resurrection, it is a fantasy to expect resurrection lead to eternal life outside of Christ.

All humans arrive at the point of death. Those who are in Christ pass through it, as Jesus has done, and those who are in the world will perish into the void of darkness when the duality dissolves. Only the good will escape the destructive clutches of duality.

According to some estimates, the last star will extinguish in 10 billion years, and the universe will become cold, dark, and silent. When the 24 billion-year history of the universe is examined retrospectively, one will see a 250,000-year window when modern humans appeared on earth, a 12,000-year period when righteous humans came into existence, a 3,500-year time when God spoke to humans, and a 33-year period in which a redeemer emerged. It is truly astonishing how narrow this window has been in the entire universe's history and how precious our redeemer is. It took light 14 billion years to redeem himself in Christ, but it only takes a moment for us to accept Jesus as our Saviour and begin the journey to eternity. It is as if the immensity and totality of all that exists converge at a precious moment to save those who are good and worthy. How fortunate are those who share redemption with the light when all else fades into darkness. A few breaths of mortality send us on a journey to immortality after repentance.

What an astonishing end to such a long and tedious struggle!

Charles Darwin understood the highly complex and destructive processes that led to life's emergence on earth. Life in all its varieties and forms struggles to evolve through natural selection or the preservation of favoured races. Darwin's work is a detailed scientific description of how living organisms have evolved in this duality: competition and elimination; preservation, accumulation, and extinction; favourable and unfavourable; advantageous and disadvantageous; divergence and convergence; monstrosity and normality. [51] There is no obvious evidence of divine involvement in these processes, and life emerges of its own accord and on its own merit. Darwin's work is undoubtedly a significant contribution to our understanding of Genesis in the Bible. Whether we accept Darwin's findings or not, the traditional biblical view of God as a creator can never remain the same. This will change our understanding of the entire biblical narrative in a way we never expected.

The great deception claims that an ostentatious temple, eye-catching religious rituals, and needless blood sacrifices are godly and lead to salvation. It is in denial of God and salvation in Jesus Christ.

We do not know how the final act will unfold on the stage of human history, but we may speculate that the Catholic Church, the Third Temple in Jerusalem, the anti-Christ, and the Islamic invasion of the Holy Land and destruction of the Third Temple will have a decisive impact on the coming events and the closing of the age.

There have been speculations and disputes about the identity of Jesus Christ. It is claimed that Jesus was a Jew. However, this cannot be so since Judaism could never have produced a Messiah who preached submission to the will of Rome. Jewish history demanded that a Messiah wage war against the might of Rome to liberate the Jews from servitude and oppression. This is not a universal Messiah.

Duality in human nature is made of: good; faith; honesty; courage; care; love; compassion; mercy; generosity; sharing; forgiveness; justice; integrity; self-control; and prudence. And bad is made of: hate; indifference; greed; dishonesty; cruelty; harm; injustice; arrogance; delusion; envy; selfishness; ego; lies; and murder. Salvation is in Jesus Christ, in whom all that is good is incarnate. So, in him, who is the bridal chamber, the soul rests, and the truth and mystery of the universe are revealed.

Jesus said, "The kingdom of God does not come in such a way as to be seen. No one will say, 'Look, here it is!' or 'There it is! ; because the kingdom of God is within you."

(Luke 17:20-21)

Within us, there is the duality of good and bad. The good takes us to the kingdom of God, and the bad takes us to hell and damnation.

From Alpha, when light was separated from darkness to produce duality

↓

To Omega, when duality was dissolved in the person of Jesus Christ to create perfect unity for all eternity

The genius of our Lord God lies in realising the power embroiled in duality when light is separated from darkness. It is in this duality that all things came into existence of their own accord over a long span of time.

His majesty is in creating a seed that, when placed in this duality, produced an Abrahamic faith that, when purified in the person of Jesus Christ, paved the way to eternal life in the kingdom of light.

And His beauty is in his fatherhood to forgive sins and grant mortals liberty from duality and eternal life in the perfect unity of Jesus Christ.

But where are our gratitude and repentance?

Have you ever wondered why the earth spins on its axis in space? As it does so, it is continuously wrapped around by day and night, which is a duality. When the spinning stops, the duality ends, and all life will come to an end, no matter how exotic, complex, and resilient it is. So where is the creativity?

What benefit is gained when a perfect seed is sown in imperfect soil? When the tree produces fruit, one wonders where the bad fruit came from.

The perfect seed that the Lord God sowed in the duality of light and darkness produced the righteous. But where do the unrighteous come from?

Chapter 5

Commentary on the Apocrypha of Philip and Thomas

The Apocrypha of Philip and Thomas provide the most fascinating reading. In this chapter, a commentary on some of the sections of the Apocrypha will be presented. This will complement the information in the Bible and show the depth of the biblical revelation as well as Jesus' sacred teachings to his disciples.

The gospel according to Philip [1]

- Whoever does good deeds in this world will benefit from the rewards in the next. It is no good to reap the rewards in this world because they are not yet mature and ready. For now, it is best to do good and be patient until the right time comes to reap the rewards. **(4)**

- Some humans have associations with truly good things. But the rulers deceived them by taking the names of the good things and giving them to the bad things. They took their freedom and enslaved them for good. **(9)**

- The truth of Jesus is everywhere. Many see it being preached, but only a few see it bearing fruit. **(13)**

- The soul, where the most precious inner faith is kept, is hidden in the worthless mortal body. **(20)**

- Our Father in heaven owns so much that belongs to his children. If we are foolish and childish at heart, they will not belong to us. But when we grow up, the Father will give them all to us. **(32)**

- When God dips the soul in the water of baptism, it soaks up its own colour and becomes imperishable. Like the truth, God's colour is imperishable. **(37)**

- The children of God are like pearls. No matter what circumstances they live in, they are always precious to God. **(41)**

- Humans travel on tediously long and laborious journeys to nowhere and find themselves staying in the same place. They experience nothing of God, and their lives are in vain. **(45)**

- There is evil and there is good in this world. But the world's goods are not really good, and its evils are not really evil. There is a world in which there are evils that are truly evil. It is death. It is preferable to obtain resurrection now so that when we die, we can find peace and rest. **(55)**

- In those days, there was oneness; there was no death. But when duality came into existence, death took over the world. When duality dissolves and unity is restored, death will disappear. **(63)**

- Unless a man is reborn by the anointed Christ, he will not receive light and water to see himself. **(67)**

- Those who put on the perfect light by joining Christ will not be seized by the spirits of wickedness. **(69)**

- Jesus came to bring what was divided back together and to give life to those who had died because of separation. Duality has caused the separation. **(70)**

- Humanity was created from light and the earth. Christ was born as a unity to rectify the duality that has caused so much suffering since the beginning. **(74)**

- Everything in the world dies. Those who live in truth are consumed by the world of truth and will not die. Jesus came from the world of truth to bring food and give it to anyone who asked for it to live. **(81)**

- It is the anointing by Christ that makes us Christians, not the water of baptism. Our father anointed Jesus, and Jesus anointed the apostles, and the apostles anointed us. How fortunate we are that the Father has given us the resurrection, light, cross, and holy spirit. Both the Father and Jesus are one, and in this oneness, all things are made perfect. **(83)**

- The light that created the world wanted it to be incorruptible and immortal. But the light had corruptibility in him, and hence the world he made became corrupt. All things are corrupt. It is only in children that incorruptibility is achieved. Adam was corrupt, but his corruptibility was corrected in his offspring, Jesus Christ. Both the light and Adam and his offspring can thus be made incorruptible in Christ. **(85)**

- When we go into the water of baptism, we draw off death. We are no longer withered away by the wind of indignation in the world but are saved by the holy spirit. **(92)**

- People who hate sin become free through the grace of God. But when they enslave themselves to sin, they cannot become free again. **(97)**

- For now, we are mesmerised by the majesty and immensity of what we see. The visible world wields enormous power over our senses. But how about hidden things? The mysteries of truth that we cannot comprehend with our senses are hidden in images and symbols. The veil conceals the truth, but when the veil is torn, the glory of the truth will destroy the falsehood, and all immortality will flee from the world. Those servants of God will enter the veil, where the hidden realm of truth rests in glory and might. Whatever there is in this realm is powerless and detestable compared with the hidden glory, but it is through it that we enter the perfect glory. The holiest of holies and the hidden aspects of truth were opened to us through Christ so that we could partake in the feast of grace and forgiveness. **(105)**

- The seed of the Holy Spirit is hidden, and evil is in its midst. When this seed is revealed, then perfect light will shine on each person, and those who belong to it will be anointed. Slaves will be set free, and captives will be ransomed. All who are separated will join to become full. **(106)**

The gospel according to Thomas (1)

● A person must not stop looking for the truth until he finds it. When he finds it, he will be disturbed and then amazed. He will then understand it in its entirety. **(2)**

● Consider what is before you, and what is incomprehensible will be disclosed to you. There is nothing hidden that will not be shown to those who seek it. **(5)**

● The disciples asked Jesus, "What do you want us to do?" Jesus said, "Do not lie, and do not do what you hate." All your actions are known to God. **(6)**

● Humans are like fishermen who cast their nets into the sea of gifts the world offers. A few find the precious gift of salvation and throw away the rest, but the majority are content with fame, fortune, and worldly pleasures. **(8)**

● Jesus came to set the world on fire, and he is watching it until it blazes. **(10)**

● The kingdom of heaven is made of humble and meek people. Humility and meekness, like a tinny mustard seed that falls on ploughed soil and grows into massive foliage, become a refuge for others. **(20)**

● Oh, faithful, be on your guard against the world. Arm yourselves with the great power of salvation in Jesus Christ. The trouble that you expect will come, and unless you are with the one who has power over it, you will all perish. **(21)**

• When the farm of faith produced its crops, the gardener came in haste, sickle in hand, and harvested them. **(21)**

• A person of light has light within them. And it illuminates the entire world. The world is full of darkness, and so little light shines. What a terrible state of affairs! **(24)**

• Those who love the word and the things of the world will never find God. It is essential to make time for God in our lives; otherwise, we will never find Him. **(27)**

• The devil cannot get to a person without first trying that person's hand. When he finds a weakness in the person's armour, he then attacks him. Reinforce your armour in the Lord Jesus Christ to keep the devil at bay. **(35)**

• A person who has good in him will be given more. A person who does not have any good in him will be deprived of the little he has. **(41)**

• Do not get involved in the affairs of this world. Be a bystander. **(42)**

• Those who feel they do not belong in this world are solitary and superior. They will find the kingdom of God because they came from it and will return to it. **(49)**

- Those who take up their cross and follow Jesus Christ must leave behind all their earthly connections and desires; otherwise, they will not be worthy of him. **(55)**

- Blessed are those who labour and find life in the person of Jesus Christ. **(58)**

- Seek the one in whom salvation lies while you are alive. A dead person cannot find the one who is alive. **(59)**

- Anyone who is accustomed to the truth but falls short in deeds and conduct falls completely. **(67)**

- Every person has light and darkness in him. The light in him will save him. But if a person has no light in him, the darkness in him will kill him. **(70)**

- So many are standing at the door of the kingdom of God, knocking and wanting to enter. Only those who are not in servitude to this world and its Satanic values will enter the kingdom. **(75)**

- Whoever is near Jesus is near the fire, and whoever is far from him is far from the kingdom. **(82)**

- The kingdom of God is like taking redemption and mixing it with the souls of the righteous to produce an ever-growing good into eternity. **(96)**

- Those who do the will of the Father will enter the kingdom of God. **(99)**

- When we leave behind the duality of this world and embrace the oneness and unity in Jesus Christ, then our faith will move mountains. **(106)**

- Whoever is acquainted with the teachings of Jesus Christ and follows them, to that person the hidden things will be made manifest. **(108)**

- The kingdom of God is spread out over the earth, and it is in the hearts of righteous humans. **(113)**

The three pillars of our existence

The scheme of Providence

In the duality of light and darkness, all things, good and bad, came into existence of their own accord. As time passed, a totality emerged in which all that was worthy was unified into one entity, the person of Jesus Christ. In him, duality was dissolved, and perfect unity arose, in which both man and God were freed from the clutches of duality. Then, the purpose of duality was defined as a struggle in which the good emerged victorious and was incarnated in Christ. The totality of Christ gained dominion in the duality of light and darkness despite immense hostility, nullifying the work of darkness and saving souls. This opened a path to eternal life and prepared the way for the redemption of humanity. Now, as duality reaches its violent end, the healing light of Christ will shine on the elect, never to be extinguished. Righteousness will be freed from the destructive stranglehold of duality, and God will prevail forever.

The tragic story of ancient Israel

In rejecting the fatherhood of God and resorting to self-reliance, ancient Israel alienated itself from the care, protection, mercy, and light of the Lord God. In demanding to become like the nations around it, Israel brought itself into permanent conflict with the gentiles. It is tragic that Israel neither became a model nation nor a light to the world as God intended it to be, nor was it accepted as an equal among the gentiles. The rejection of Jesus Christ as its saviour sealed the fate of ancient Israel for perpetuity.

So, what has destiny stored for Israel in the future? Light or darkness?

The Kingdom of Heaven

The Kingdom of Heaven is where the best of all humans who have ever lived will be united into a single nation under the fatherhood of God. It is where light will reach its ultimate purity and express itself in its true colours: peace, justice, forgiveness, healing, caring, loving, giving, joy, and companionship.

This is where Jesus says, "I am."

References

1 – B. Layton. The Gnostic Scriptures. SCM Press Ltd, London, 1987. (ISBN: 0-334-02022-0).

2 – J. B. Lumpkin. The Books of Enoch, The angels, the watchers and the Nephilim. Fifth Estate Publishers USA, 2011. (ISBN: 9781936533077).

3 – https://en.wikipedia.org/wiki/Galaxy   Date visited: 14-01-2023

4 – Voyage Through the Universe. Galaxies. Arcata Graphics, U.S.A., 1992. (ISBN: 0 7054 1070 6).

5 – L. M. Krauss. A Universe from Nothing. Simon & Schuster UK, London, 2012. (ISBN: 978-1-147111-268-3)

6 – https://en.wikipedia.org/wiki/Ultimate_fate_of_the_universe
Date visited: 14-01-2023

7 – https://en.wikipedia.org/wiki/Formation_and_evolution_of_the_Solar_System Date visited: 26-01-2023

8 – https://www.space.com/19321-sun-formation.html
Date visited: 26-01-2023

9 – https://cneos.jpl.nasa.gov/about/life_on_earth.html
Date visited: 27-01-2023

10 – https://en.wikipedia.org/wiki/Evolutionary_history_of_plants
Date visited: 28-01-2023

11 – https://sci.waikato.ac.nz/evolution/AnimalEvolution.shtml
Date visited: 28-01-2023

12 – https://en.wikipedia.org/wiki/Evolution_of_reptiles
Date visited: 30-01-2023

13 – https://en.wikipedia.org/wiki/Mammal
Date visited: 30-01-2023

14 – https://en.wikipedia.org/wiki/Extinction_event
Date visited: 30-01-2023

15 – E. Peter Volpe and P A. Rosenbaum. Understanding evolution. McGraw Hill Higher Education. 2000. (ISBN: 0-697-05137-4)

16 – https://www.google.com/search?client=firefox-b-d&q=what+is+hominid%3F   Date visited: 31-01-2023

17 – H. Wendt. From Ape to Adam. The search for the ancestry of man. Book Club Associates. London. 1974.

18 – N. Smart. The World's Religions. Cambridge University Press, UK. 1998. (ISBN: 0 521 63748 1).

19 – https://en.wikipedia.org/wiki/List_of_empires
Date visited: 08-02-2023

20 – https://www.vaticancitytours.it/blog/the-rise-and-fall-of-the-roman-empire/
Date visited: 08-02-2023

21 – https://en.wikipedia.org/wiki/Roman_Empire
Date visited: 08-02-2023

22 – https://rome.mrdonn.org/achievements.html
Date visited: 080-02-2023

23 – S. Baker. Ancient Rome. The Rise and Fall of an Empire. BBC Books, 2007. (ISBN: 978 1 846 07284 0).

24 – https://en.wikipedia.org/wiki/Battle_of_the_Milvian_Bridge
Date visited: 08-02-2023

25 – https://en.wikipedia.org/wiki/Constantine_the_Great_and_Christianity
Date visited: 08-02-2023

26 – R. Montgomery and B. O'Dell. The List. Persecution of Jews by Christians throughout history. Root Source Press. Israel. 2019. (ISBN: 978-965-7738-13-9)

27 – https://en.wikipedia.org/wiki/Napoleon_and_the_Catholic _Church
Date visited: 09-02-2023

28 – https://en.wikipedia.org/wiki/Vatican_City
Date visited: 09-02-2023

29 – https://webwitness.org.au/estimates.html
Date visited: 09-02-2023

30 – https://en.wikipedia.org/wiki/European_wars_of_religion
Date visited: 09-02-2023

31 – https://en.wikipedia.org/wiki/Age_of_Discovery
Date visited: 10-02-2023

32 – https://en.wikipedia.org/wiki/European_balance_of_power
Date visited: 10-02-2023

33 – https://www.jewishgen.org/databases/givennames/enlitmnt.htm
Date visited: 11-02-2023

34 – https://www.smithsonianmag.com/smart-news/researchers-find-evidence-pope-pius-xii-ignored-reports-holocaust-180974795/
Date visited: 11-02-2023

35 – https://en.wikipedia.org/wiki/History_of_colonialism
Date visited: 10-02-2023

36 – https://en.wikipedia.org/wiki/Anti-Zionism
Date visited: 15-02-2023

37 – https://en.wikipedia.org/wiki/Balfour_Declaration
Date visited: 10-02-2023

38 – https://en.wikipedia.org/wiki/History_of_Zionism
Date visited:10-02-2023

39 – https://www.commentary.org/articles/fe-cartus/vatican-ii-the-jews/
Date visited: 12 -02-2023

40 – https://www.jewishvirtuallibrary.org/christian-zionism
Date visited: 1202-2023

41 – https://free.messianicbible.com/feature/israels-priests-prepare-third-temple/
Date visited: 10-20-2023

42 – https://www.jewishgen.org/databases/givennames/enlitmnt.htm
Date visited: 11-02-2023

43 – https://www.voanews.com/a/pope-justice/1625457.html
Date visited: 10-02-2023

44 – https://jewsforjudaism.org/knowledge/articles/why-is-israel-targeted
Date visited: 13-03-2023

45 - https://www.weforum.org/about/world-economic-forum
Date visited: 12-02-2023

46 – https://now.tufts.edu/2022/11/18/digital-dollar-coming-soon
Date visited: 12-02-2023

47 – https://agriculture.vic.gov.au/farm-management/soil/what-is-soil
Date visited: 14-02-2023

48 – https://blog.pedersen-group.co.uk/blog/what-is-poor-soil-quality-and-how-can-it-impact-your-crops
Date visited: 14-02-2023

49 – https://www.iamcountryside.com/growing/soil-health-what-makes-good-soil/
Date visited: 14-02-2023

50 – https://www.sciencedirect.com/science/article/abs/pii/S0889157520313168

51 – C. Darwin. The origin of species. Avenel Books. New York. 1979. (ISBN: 0-517-30978-5)

All scripture quotations have been taken from the Interlinear Bible, Hebrew-Greek- English. Jay P. Green, Sr Hendrickson Publishers, 2020. (ISBN: 978-1-56563-977-5) 8.

Afterword

The narrative in the Bible has been the subject of scholarly investigation and debate since it was made available to the common man. A narrative must be consistent, beneficial to individual faith, and relevant in its content. Faith is by nature personal, and the faithful understand the Bible in different ways. We weak mortals can only struggle to understand God and the immensity of his created work. Let's hope the story in this book, as little as it may be, will benefit its readers to the glory of God and our Lord Jesus Christ. We can add only a drop to this vast pool of knowledge and never stop wondering in amazement at its complexity and majesty. Despite major advances made in cosmology in recent years, our understanding of the exact events that led to the creation of the universe remains incomplete. Still, we must journey into our imaginations and think of a scenario in which the universe was made.

About the author

Dr. Ali Ansarifar has been living in the U.K. for over 40 years. He was awarded a bachelor's degree and a doctorate in Materials Science from Queen Mary College, the University of London, and a Diploma in Interface Science from Imperial College, the University of London. He worked as a post-doctoral research assistant at Imperial College, London, and the Cavendish Laboratory, Department of Physics, University of Cambridge. He was an upper senior research scientist in a rubber research and development centre in Hertfordshire, U.K., and a lecturer in Polymer Engineering in the Materials Department at Loughborough University until he retired as a senior lecturer. He has given lectures, seminars, and workshops in the United States, the United Kingdom, Europe, the Middle East, and Southeast Asia; published over 150 technical research papers in peer-reviewed international scientific journals and technical magazines for the polymer and tire industries and textbooks; and contributed chapters to scientific books. He has been on the editorial boards of rubber and adhesion scientific journals and has been awarded prizes for his scientific publications. He is a Fellow of the Higher Education Academy, U.K., and a servant of Jesus Christ.

About the book

This book briefly discusses the following topics: the origin of the universe, the planetary system, the Earth, and life on Earth; the impact of Christianity on Western European history and humanity; the rise of the Papacy and the challenges it has been facing since the Protestant Reformation. The Counter-Reformation's master plan to restore the credibility of the Popes and the institutions of the Catholic Church through deception will be discussed, and some ideas will be put forward about the unfolding events and the rise of the New World Order. The book also examines the biblical creation narrative in Genesis and the image of God as the Creator, and it includes a short commentary on the apocrypha of Philip and Thomas, which are not included in the Bible. It is hoped that this book will help readers appreciate the importance of salvation in Jesus Christ and take advantage of this short moment of grace in the entire history of the world before it all ends.

Final notes by the author

The debate about Jesus Christ and his identity has been raging on for almost two thousand years. Jesus Christ has never been resurrected. He is still buried under historical claims and counterclaims, controversies, doubts, and denials. Men will never reach an agreement about the real identity of Jesus Christ and his place in the scheme of human existence and experience. It is left to the Bible to do it.

The verses in Genesis 1–5 set the scene for the entirety of the biblical narrative. These verses suggest that all things exist in a medium of duality that started when light and darkness were separated. This medium is imperfect but highly creative. It produces very bad things and amazingly good things through immense struggle and destruction over an unimaginable span of time. The bad and the good must be separated permanently at the end to preserve the good for perpetuity. The Christian Gospel declares that God has achieved the final separation of good and evil in Jesus Christ, and in him rests the perfect unity of all that is good and precious in human nature. No human wisdom is needed; only trust in the word of God and accept the salvation offered in Jesus Christ. This will free us from the imperfect duality and its dreadful end. Repentance is the first step to freeing oneself from the clutches of duality, and then, by God's mercy, comes the union with the perfect unity of Christ.

Books by the author

1 – Sharing the Faith. Kingdom Publishers, London 2021. ISBN: 978-1-913247-54-6

2 – The Bible Story of Mankind. A Covenant with God with no Get-out Clause. Balboa Press, UK 2021. ISBN: 978-1-9822-8394-0 (sc), ISBN: 978-1-9822-8393-3 (e).

3 – Why Did God Create Mankind? The Problem of Duality with God. Balboa Press, UK 2021. ISBN: 978-1-9822-8439-8 (sc), ISBN: 978-1-9822-8440-4 (e).

4 – The British and American Empires and the State of Israel. Until the Kingdom of God Comes. Kingdom Publishers, London 2022. ISBN: 978-1-913247-98-0

5 – The March to the Armageddon, Balboa Press, UK 2022. ISBN: 978-1-9822-8377-3 (sc), ISBN: 978-1-9822-8378-0 (e)

6 – How Did God Create Mankind? Scientific and Biblical Views. Kingdom Publishers, UK 2022. ISBN: 978-1-911697-64-0

7 – Why does Judaism reject Jesus Christ? Unintended Messianic expectations and unplanned temple-worship. Kingdom Publishers, UK 2023. ISBN: 978-1-911697-73-2

www.ingramcontent.com/pod-product-compliance
Lightning Source LLC
Chambersburg PA
CBHW041147110526
44590CB00027B/4149